ACA
Common Core Mathematics Teacher's Manual
Grades 2 & 3

for

Aesop's Childhood Adventures
Series

Vincent A. Mastro

©2014 *Vangelo Media*

ACA Common Core Mathematics Teacher's Manual
Grades 2 & 3
for Aesop's Childhood Adventures Series

Copyright © 2014 by Vincent A. Mastro. All rights reserved.

Except as permitted under the United States Copyright Act, no part of this publication may be reproduced or distributed in any form or by any means, or stored in a database retrieval system, without prior written permission of the publisher.

First published 2013 by *Vangelo Media*
Special discounts are available on quantity purchases. For details, send inquiries to info@vangelomedia.com or visit www.vangelomedia.com.

Printed in the United States of America

Publisher's Cataloging-in-Publication data

Mastro, Vincent.
   ACA common core mathematics teacher's manual: Grades 2 & 3 for Aesop's childhood adventures / Vincent A. Mastro.
   p. cm.
   ISBN: 978-1-940604-25-1

Audience: educators.
Summary: Pedagogical guidance on applying the Common Core standards to the stories of Aesop's Childhood Adventures Series.
Contents: The tortoise and the hare; The crow and the pitcher; Common core standards MD.2 and MD.3; common core mapping; and worksheets.

1. xxx
2. Aesop's fables—Adaptations. [1. Fables. 2. Folklore.] I. Aesop. II. Title.

Table of Contents

Introduction	**5**
Common Core Mapping Summary	**6**
The Tortoise and the Hare	**9**
Story	10
Common Core Standard mapping for Mathematics: Measurement & Data - Grade 2	22
Worksheets	23
The Crow and the Pitcher	**30**
Story	31
Common Core Standard mapping for Mathematics: Measurement & Data - Grade 3	40
Worksheets	41

Introduction

This teacher's manual provides pedagogical guidance on applying the Common Core, mathematics standards to *The Tortoise and the Hare* and *The Crow and the Pitcher* stories of Aesop's Childhood Adventures Series. The purpose is to facilitate the teacher's efforts to create practical lesson plans. This guidance is provided in the form of tables that contain both the core standard and the guidance as illustrated below:

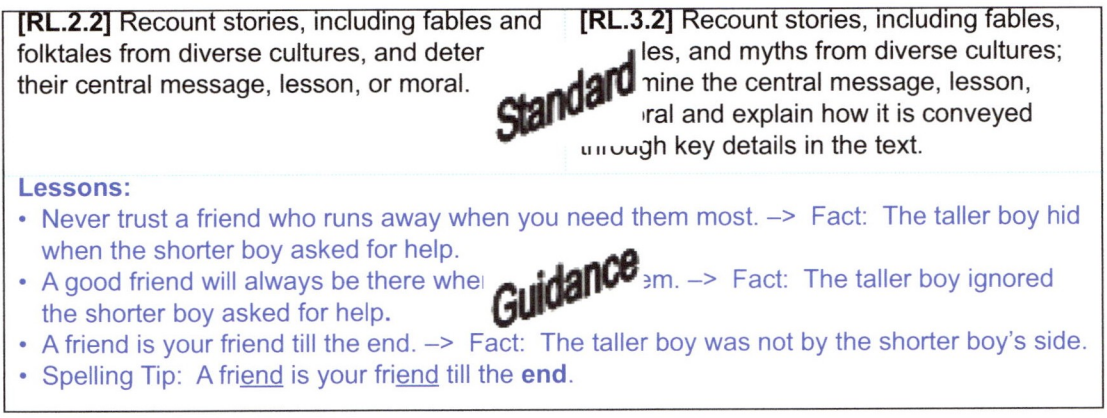

This manual also includes worksheets and the pages of each story.

Spiral bound: The print version of this guide was formatted so that it could be spiral bound. Any printer (such as Kinko's) can do this for a reasonable fee. They will cut the binding, punch the holes and add the spiral binding.

We are always looking for comments, fixes, updates and enhancement. Please contact us at anytime with your suggestions at: info@vangelomedia.com or go to the contact page of our website: www.vangelomedia.com.

ACA Common Core Mathematics Teacher's Manual

Aesop's Childhood Adventures Series
Common Core Mapping
Summary

Common Core Standard Reading Literature		Tortoise Hare	Friends Bear	Crow Pitcher	Goose Egg	Equal Shares	Oak Grass
Grade 2	CCSS.ELA-Literacy.RL.2.1	X	X	X	X	X	X
	CCSS.ELA-Literacy.RL.2.2	X	X	X	X	X	X
	CCSS.ELA-Literacy.RL.2.3	X	X	X	X	X	X
	CCSS.ELA-Literacy.RL.2.4	X	X	X	X	X	X
	CCSS.ELA-Literacy.RL.2.5	X	X	X	X	X	X
	CCSS.ELA-Literacy.RL.2.6	X	X	X	X	X	X
	CCSS.ELA-Literacy.RL.2.7	X	X	X	X	X	X
	CCSS.ELA-Literacy.RL.2.9	X	X	X	X	X	X
Grade 3	CCSS.ELA-Literacy.RL.3.1	X	X	X	X	X	X
	CCSS.ELA-Literacy.RL.3.2	X	X	X	X	X	X
	CCSS.ELA-Literacy.RL.3.3	X	X	X	X	X	X
	CCSS.ELA-Literacy.RL.3.4	X	X	X	X	X	X
	CCSS.ELA-Literacy.RL.3.5	X	X	X	X	X	X
	CCSS.ELA-Literacy.RL.3.6	X	X	X	X	X	X
	CCSS.ELA-Literacy.RL.3.7	X	X	X	X	X	X
	CCSS.ELA-Literacy.RL.3.9	X	X	X	X	X	X

Common Core Standard Reading Informational Text	Tortoise Hare	Friends Bear	Crow Pitcher	Goose Egg	Equal Shares	Oak Grass
Grade 2 CCSS.ELA-Literacy.RI.2.1	X	X	X	X	X	X
CCSS.ELA-Literacy.RI.2.2	X	X	X	X	X	X
CCSS.ELA-Literacy.RI.2.3	X		X			
CCSS.ELA-Literacy.RI.2.4	X	X	X	X	X	X
CCSS.ELA-Literacy.RI.2.5	X	X	X			
CCSS.ELA-Literacy.RI.2.6	X	X	X	X	X	X
CCSS.ELA-Literacy.RI.2.7	X	X	X			
CCSS.ELA-Literacy.RI.2.9	X	X	X	X	X	X
Grade 3 CCSS.ELA-Literacy.RI.3.1	X	X	X	X	X	X
CCSS.ELA-Literacy.RI.3.2	X	X	X	X	X	X
CCSS.ELA-Literacy.RI.3.3	X		X			
CCSS.ELA-Literacy.RI.3.4	X	X	X	X	X	X
CCSS.ELA-Literacy.RI.3.6	X	X	X	X	X	X
CCSS.ELA-Literacy.RI.3.7	X	X	X			
CCSS.ELA-Literacy.RI.3.9	X	X	X	X	X	X

Common Core Standard Mathematics: Measurement & Data	Tortoise Hare	Friends Bear	Crow Pitcher	Goose Egg	Equal Shares	Oak Grass
Grade 2 CCSS.Math.Content.2.MD.A.1	X					
CCSS.Math.Content.2.MD.A.2	X					
CCSS.Math.Content.2.MD.A.3	X					
CCSS.Math.Content.2.MD.A.4	X					
CCSS.Math.Content.2.MD.D.9	X					
Grade 3 CCSS.Math.Content.3.MD.A.2			X			

Common Core Standard Writing		Tortoise Hare	Friends Bear	Crow Pitcher	Goose Egg	Equal Shares	Oak Grass
Grade 1	CCSS.ELA-Literacy.W.1.1	X	X	X	X	X	X
	CCSS.ELA-Literacy.W.1.2	X	X	X	X	X	X
	CCSS.ELA-Literacy.W.1.3	X	X	X	X	X	X
	CCSS.ELA-Literacy.W.1.5	X	X	X	X	X	X
	CCSS.ELA-Literacy.W.1.6	X	X	X	X	X	X
	CCSS.ELA-Literacy.W.1.8	X	X	X	X	X	X
Grade 2	CCSS.ELA-Literacy.W.2.1	X	X	X	X	X	X
	CCSS.ELA-Literacy.W.2.2	X	X	X	X	X	X
	CCSS.ELA-Literacy.W.2.3	X	X	X	X	X	X
	CCSS.ELA-Literacy.W.2.5	X	X	X	X	X	X
	CCSS.ELA-Literacy.W.2.6	X	X	X	X	X	X
	CCSS.ELA-Literacy.W.2.8	X	X	X	X	X	X
Grade 3	CCSS.ELA-Literacy.W.3.1	X	X	X	X	X	X
Grade 4	CCSS.ELA-Literacy.W.4.1	X	X	X	X	X	X

The Tortoise and the Hare

Common Core
Mapping

The Tortoise and the Hare

"But why Nana, why?" asked Aesop.

With a big smile and a sparkle in her eye, the older one said, "Aesop, I don't know the answer to every question. Sometimes, you have to go and find the answer for yourself."

"I will, Nana. Today is the day I will find out why."

Little Aesop looked up at his grandmother. He jumped

out of her lap and on to the floor. He ran through the den, down the hall and out the door.

It was a beautiful sunny day, perfect weather for Little Aesop to go out and play.

Soon, Little Aesop came upon his friends the hare, the fox, the two mice, and the tortoise.

"Hi everyone," said Aesop.

The hare turned to Aesop and said, "I was just saying that I can run so fast that I can beat anyone in a race."

The tortoise shook her head sadly and asked, "Why do you need to brag so much?"

"It is not bragging when it is true," said the hare.

The other animals looked at each other and said nothing.

The hare turned toward the tortoise and laughed at her. "Look how short your legs are. You must be the slowest of us all!"

Aesop was disappointed in the hare for being so rude.

The tortoise replied, "I may be slow and have short legs, but I have something that you do not have."

The hare chuckled and said, "That is right, Tortoise. You do have something that none of us have. You have a big heavy shell to slow you down."

"There you go boasting again," said the tortoise.

"Do you think you can beat me in a race?"

The hare burst into laughter, as did all the other animals, except for Aesop. Aesop was worried that the tortoise would lose the race. He did not want her to be sad and he knew the hare would never stop bragging if he won.

"Is that a joke?" asked the hare. "I could dance the whole way and still beat someone as slow as you."

The mouse jumped up and down with excitement and yelled, "A race, a race! I'm going to get my checkered flag."

When the mouse returned he stood by the tree and said, "The race will start here. You will run past the big boulder, and go down the hill. Then you will run around the clover field and back up the hill to this tree. The first one to the finish line wins the race."

The tortoise was ready at the starting line, while the hare was gibbering and jabbering with all the other animals.

Aesop was worried that the race was much too long for the tortoise. He did not want her to lose.

"On your mark … get set … GO!" yelled the mouse.

The tortoise pushed herself forward, taking one slow step at a time. She was moving as fast as her little legs could take her, but she did not get very far. All the other animals pointed at her and laughed, except for Aesop.

The hare was laughing too. They laughed until their bellies ached.

The mouse saw that the hare had not yet started the race so he yelled, "Go hare, go!"

Swoosh!

The hare took off like a dart. He ran past the tortoise, by the boulder, then down the hill, and was out of sight.

The tortoise had not gone very far at all. She had not even reached the big boulder. One of the animals yelled, "You should just give up, Tortoise. There is no way you are going to win. You are much too slow."

Aesop called out to the tortoise, "Go, Tortoise, go! You can do it!"

The tortoise looked over and smiled at Aesop and said, "I will never give up. I have something that the hare does not have." But no one understood what the tortoise was talking about.

Aesop and all the other animals ran to the top of the hill to see how far the hare had gone. To everyone's surprise, the hare was lying down next to a stump playing with the clover.

The hare saw the animals and yelled, "Look at all this clover. I am going to find one with four leaves."

After a while, the hare got bored looking for a four leaf clover and he fell asleep.

Aesop could not believe that the hare had fallen asleep during the middle of a race. It was at that moment, that Aesop understood what the tortoise meant when she had said, that she had something that the hare did not have.

Aesop smiled and walked to the finish line because he now knew, the tortoise would win the race.

Meanwhile, the tortoise just plodded on, taking one slow step at a time, focusing only on the finish line. Eventually, the tortoise passed the sleeping hare, who was snoring as he napped in the clover. She continued slowly on her way.

When the hare woke up from his nap, he saw that the tortoise was just about to cross the finish line. He leapt up, and ran as fast as he could.

But he was too late. The tortoise had crossed the finish line first and won the race!

"Congratulations!" said the fox. "It looks like slow and steady wins the race."

"That is right," replied the tortoise. "I have perseverance."

"What is perseverance?" asked the fox.

Aesop knew that the tortoise won the race because she had focused only on the race and did not give up.

Little Aesop could not wait to go home and tell his Nana about perseverance. He said good bye to his friends and headed home.

"I'm happy to see you my little one," said Aesop's grandmother, who was sitting in the den. "Did you find what you were looking for? Did you get the answer to your questions?"

"No," said Aesop as he climbed into his grandmother's lap, "I still do not know why, but I did learn that slow and steady can win a race, and I don't like it when people boast."

"That is very good Aesop, tell me more."

"Well, I saw a race today, Nana, where the tortoise beat the hare…"

When Aesop was finished telling his grandmother everything that happened, she looked at the young one with a big smile and a sparkle in her eye and said, "It sounds like you had a great adventure today. I know you did not find what you were looking for, but you did learn that bragging can hurt people and anything is possible with perseverance."

She then hugged Aesop and said, "Don't worry about your questions, Aesop. Eventually, you will find the answers to all of them because you have perseverance too."

ACA Common Core Mathematics   Teacher's Manual

The Tortoise and the Hare   Common Core Mapping   Mathematics: Measurement and Data

Grade 2
Measure and estimate lengths in standard units
[2.MD.A.1] Measure the length of an object by selecting and using appropriate tools such as rulers, yardsticks, meter sticks, and measuring tapes.
1. Have the children group up and draw a picture of the race track based on the description in the story (see the *What does the race track look like?* worksheet.) 2. Give each group some string and have them lay the string along the entire route and cut the string to match the length of the race track route. 3. Stick the pictures of the race track on the wall, side by side. 4. Ask each group to write down which race track they think is the longest. 5. Have each group measure their string and write the length down on a sticky. 6. Put the sticky on the appropriate race track. 7. Discuss why the longest is the longest and the shortest is the shortest. NOTE: To help with this discussion you may want to create two race tracks that are the exact same length and height. One of them will be shaped like an oval the other shaped like a figure eight. Cut a string for each track and compare and contrast them.
[2.MD.A.2] Measure the length of an object twice, using length units of different lengths for the two measurements; describe how the two measurements relate to the size of the unit chosen.
Have each group measure the length of their string twice, using length units of different lengths for the two measurements; describe how the two measurements relate to the size of the unit chosen. (see *Estimate and measure with different units* worksheet.)
[2.MD.A.3] Estimate lengths using units of inches, feet, centimeters or meters.
Using the string, estimate lengths using units of inches, feet, centimeters, and meters (see *Estimate and measure with different units* worksheet.).
[2.MD.A.4] Measure to determine how much longer one object is than another, expressing the length difference in terms of a standard length unit.
1. Using the lengths of the strings from each group, determine how much longer one string is than another, expressing the length difference in terms of a standard length unit. Document this in the *Which track is longer* worksheet. 2. Place the longest string on the shortest track to illustrate the difference.
Represent and interpret data
[2.MD.D.9] Generate measurement data by measuring lengths of several objects to the nearest whole unit, or by making repeated measurements of the same object. Show the measurements by making a line plot, where the horizontal scale is marked off in whole-number units.
Using the measurements from each group, generate measurement data. Show the measurements by making a line plot, where the horizontal scale is marked off in whole-number units (see *Plot a line graph* worksheet.)

©2013 *Vangelo Media*

Common Core Mathematics Worksheets

for

The Tortoise and the Hare

What does the race track look like?

"The race will start here. You will run past the big boulder, and go down the hill. Then you will run around the clover field and back up the hill to this tree. The first one to the finish line wins the race."

Name: _____ Date: _____

Estimate and measure with different units.

Length units of measure:
inch, foot, yard, mile, league, centimeter, meter, kilometer, hand, light-year, ...

	_____ Units	_____ Units
1st measurement		
2nd measurement		

1st Estimate _____ Units

2nd Estimate _____ Units

Name: _____ Date: _____

Which track is longer?

Length units of measure:
inch, foot, yard, mile, league, centimeter, meter, kilometer, hand, light-year, ...

_____
Units

1st Track Length

2nd Track Length

Difference

Name: _____ Date: _____

ACA Common Core Mathematics   Teacher's Manual
Plot a line graph.

_____
   Units

Length

Track 1   Track 2   Track 3   Track 4   Track 5   Track 6

Name: _____ Date: _____

How long is this race track?

_____
Length

_____
Units

Name: _____ Date: _____

How long is this race track?

_____
Length

_____
Units

Name: _____ Date: _____

The Crow and the Pitcher

Common Core
Mapping

The Crow and the Pitcher

"But why Nana, why?" asked Aesop.

With a big smile and a sparkle in her eye, the older one said, "Aesop, I don't know the answer to every question. Sometimes, you have to go and find the answer for yourself."

"I will, Nana. Today is the day I will find out why."

Little Aesop looked up at his grandmother. He jumped

out of her lap and on to the floor. He ran through the den, down the hall and out the door.

It was a beautiful sunny day; perfect weather for Little Aesop to go out and play.

Soon, Little Aesop came upon a picnic table that had a pitcher on it. He noticed that the pitcher was half full of water.

Aesop turned to continue on his way when his friend the crow came swooping down and landed next to him.

"Hi," said Aesop. "How are you? It has been a long time since I last saw you."

"Hi Aesop," whispered the crow. "I am so thirsty that I can barely talk. Do you know where I can get some water?"

"Yup, from the pitcher on the picnic table."

The bird hopped up onto the pitcher and tried to drink the water, but he could not reach it. He tried and he tried, but the water was simply not high enough.

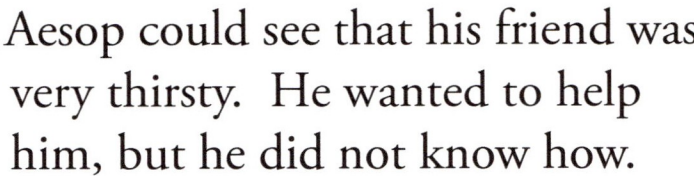

Aesop could see that his friend was very thirsty. He wanted to help him, but he did not know how.

Aesop watched as the crow strutted back and forth looking into the pitcher and then looking all around. The bird was very thirsty and very frustrated.

All of a sudden, the crow stopped strutting. He jumped off the picnic table and picked up a pebble.

"No! Stop! Don't eat that pebble!" yelled Aesop.

"Eat the pebble? I am not going to do that!" said the crow. "I am going to drop the pebble into the pitcher."

"What? Why?" asked Aesop.

"So I can drink the water. I think I can solve this problem."

"Oh boy, now I'm really confused," said Aesop.

The crow flew up onto the table, dropped the pebble into the pitcher and said, "Watch this."

Aesop was still very confused.

The bird spied another pebble, flew down and grabbed it.

He dropped that pebble into the pitcher. He then found another, and another.

The crow turned toward Aesop and said, "Look at the water in the pitcher. It has moved up a little bit hasn't it?"

"It has! I don't believe it," said Aesop. "But it is not high enough for you to drink yet. What are you going to do?"

"That is OK," said the crow. "With a little hard work, I will collect enough pebbles to get a drink. Will you help me collect pebbles?"

"I sure will," said Aesop. He was happy that there was something he could do to help his friend.

Aesop and the crow continued gathering pebbles and dropping them into the pitcher.

After a while the water rose high enough for the crow to get a drink.

When the crow was finished drinking, he said, "That was the best drink of water I have ever had."

"I'll bet it was," said Aesop.

The crow then stretched his wings and gave a big yawn. "It is time for me to get back to my nest, Aesop. Good bye, and thanks for all of your help."

"You're welcome," said Aesop as he headed home.

"I'm happy to see you my little one," said Aesop's grandmother, who was sitting in the den. "Did you find what you were looking for? Did you get the answer to your questions?"

"No," said Aesop as he climbed into his grandmother's lap, "I still do not know why, but I did learn that most problems can be fixed with a good idea and a little hard work."

"That is very good Aesop; tell me more."

"Well, I saw my friend, Crow, today. He was very thirsty…"

When Aesop was finished telling his grandmother everything that happened, she looked at the young one with a big smile and a sparkle in her eye and said, "It sounds like you had a great adventure today. I know you did not find what you were looking for, but you did learn the importance of thinking about a problem and working hard to solve it."

She then hugged Aesop and said, "Don't worry about your questions, Aesop. Eventually, you will find the answers to all of them because you have perseverance."

The Crow and the Pitcher — Common Core Mapping — Mathematics: Measurement and Data

Grade 3

Solve problems involving measurement and estimation

[3.MD.A.2] Measure and estimate liquid volumes and masses of objects using standard units of grams (g), kilograms (kg), and liters (l). Add, subtract, multiply, or divide to solve one-step word problems involving masses or volumes that are given in the same units, e.g., by using drawings (such as a beaker with a measurement scale) to represent the problem.

Objective: Determine which container holds the most water using the same method as Crow.

Materials:
- 2 or more different shaped containers that hold the exact same amount of liquid. [Do not tell the class they have the same volume. If you cannot find containers that hold the same volume, you can draw a line on the containers for the same maximum volume and tell the class that is the point where the crow can get his drink.]
- Uniformly sized marbles or pebbles.
- Measuring cup/beaker

Steps:
1. Present the empty containers to the class and ask which container holds the most water. Enter this in the Summary Data Sheet and have the children document these answers on page 1 of the "Estimate which container holds the most water." worksheet.
2. Fill the containers with the same amount of water. It should be about ¾ full (do not let the class know it is the same volume).
3. Split the class giving each group a container and pebbles.
4. Ask the children to estimate how many pebbles are required to make the water overflow by gathering them into a pile.
5. Have the children put the pebbles on page 1 of the "Estimate which container holds the most water", worksheet and flatten them so the children can draw an accurate outline of the pebbles.
6. Compare the outlines and ask which is bigger and why they differ (it is rare for them to be the same). Compare it with their previous answer to #1 above. Ask the children which container needs the most pebbles to make it overflow? Ask the children document their answer on page 1 of their worksheets and also document it on page 1 of the Summary Data Sheet.
7. Have the children slowly drop the pebbles into their container, right up to the point of spillage (or the line).
8. Have the children carefully pour the water into a bucket separating the pebbles from the water.
9. Dry the pebbles then draw an outline around them. Ask the children which container holds the most water and document their answer on page 2 of their worksheet. Have their answers changed, why?
10. Compare and contrast the pebble outlines from the groups. They should be almost the same give or take a few pebbles. Ask the children which container used the most pebbles and document their answers on page 1 of the Summary Data Sheet.
11. Ask the class how much water was in each container. (Answer: the same amount because the amount of pebbles are the same,)
12. Now have them pour out the water and measure it. Document the volume on page 3, "Measure water and pebble volume" and page 2 of the Summary Data Sheet.
13. Take the pebbles from step 11 and put them in a measuring cup. Document it on page 3, "Measure water and pebble volume" and page 2 of the Summary Data Sheet.
14. Add the water volume to the pebble volume to get the total volume. Document it on page 2 and page 3.
15. Tell the class the actual volume and ask why it differs. The children's answer will usually be larger than actual because the pebbles are round not square so there are gaps between the pebbles giving them the appearance of being a larger volume when measured.

Common Core Mathematics Worksheets

for

The Crow and the Pitcher

ACA Common Core Mathematics

Summary data sheet for
The Crow and the Pitcher

Teacher's Manual

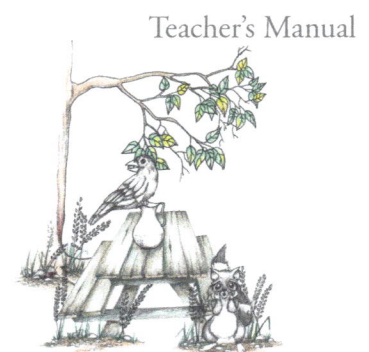

A	B	C	Same	
☐	☐	☐	☐	**Which container holds the most water?** (step 1)
☐	☐	☐	☐	**Which outline of the pebbles is bigger?** (step 6)
☐	☐	☐	☐	**Which container needs the most pebbles to make it overflow?** (step 6)
☐	☐	☐	☐	**Which outline of the pebbles is bigger?** (step 9)
☐	☐	☐	☐	**Which container needs the most pebbles to make it overflow?** (step 10)

Name: _____ **Date:** _____

Common Core Standard: 3.MD.A.2   Copyright 2013 by Vangelo Media

Summary Data Sheet

Volume units of measure:
Quarts (qts), cups (c), ounces (oz), tablespoons (tbs), teaspoons (tsp), liters (l), milliliters (ml), ...

Page 2 of 2

	Containers		
_____ Units	A	B	C
How much water is in the container? (step 12)			
What volume of pebbles were needed to fill the container? (step 13)			
What is the volume of the container? (step 14)			

Which container holds the most water? (step 15) _____

Name: _____ Date: _____

Estimate which container holds the most water.

1. Which container holds the most water?

2. Draw an outline around the pebbles needed to fill the container.

3. Which container needs the most pebbles to make it overflow? _____

Name: _____ Date: _____

Determine which container holds the most water.

4. Draw an outline around the pebbles that filled the container.

5. Which container holds the most water? _____

Name: _____ Date: _____

ACA Common Core Mathematics

Measure water and pebble volume

Volume units of measure:
Quarts (qts), cups (c), ounces (oz), tablespoons (tbs), teaspoons (tsp), liters (l), milliliters (ml), ...

Teacher's Manual

Page 3 of 3

_____
Units

How much water is in the container?

What volume of pebbles were needed to fill the container?

What is the volume of the container?

Name: _____ Date: _____

www.ingramcontent.com/pod-product-compliance
Lightning Source LLC
Chambersburg PA
CBHW041530220426
43671CB00002B/38